ESPIONAGE BLACK BOOK TWO

In this series:

Espionage Black Book One: Intelligence Databases Explained

ESPIONAGE BLACK BOOK TWO:

Codes and Ciphers Explained

Henry W. Prunckun

Bibliologica Press

Espionage Black Book Two:
Codes and Ciphers Explained

ISBN 978-0-6450643-9-1

A catalogue record for this
book is available from the
National Library of Australia

Web: bibliologica.com

Bibliologica Press
P.O. Box 656
Unley, South Australia, 5061
Australia

CONTENTS

The Zodiac Code 1

World of Codes and Ciphers 10

Developing Codes and Ciphers 23

Breaking Codes and Ciphers 33

Quest for the Unbreakable 47

An Unbreakable Cipher 54

About the Author 64

Index 65

GLOSSARY OF TERMS

Cipher—an algorithm that converts plaintext to ciphertext.

Ciphertext—a coded plaintext message.

Codebreaking—cryptanalysis.

Cracking—cryptanalysis.

Cryptanalysis—the mathematical process of deciphering messages when the key is not known.

Cryptography—the study of the principles, methods, procedures, and tradecraft used to encrypt information.

Cryptology—the academic field comprising cryptography and cryptanalysis.

Decipher—converting ciphertext to plaintext.

Decrypt—deciphering.

Decryption algorithm—the mathematical process involved in taking a ciphertext and its decryption key to produce plaintext.

Encipher—converting plaintext to ciphertext.

Encrypt—enciphering.

Encryption algorithm—the mathematical process involved in taking a plaintext message and its decryption key to produce ciphertext.

Key—information that is used to encipher and decipher secret messages, which is only known to the sender and receiver.

Plaintext—the original unencrypted message.

— CHAPTER ONE —

THE ZODIAC CODE

"Dear Editor This is the Zodiac speaking." That was the salutation of a letter sent to *The San Francisco Examiner* on August 7, 1969. The writer was the serial killer attributed with killing five people but claimed, and was suspected by police, of killing more.

On August 1, 1969, the publisher of the morning *Vallejo Herald-Times* and the evening *News-Chronicle* received a letter that contained a cryptogram that comprised letters and symbols. The characters were arranged in eight rows of seventeen characters each. There was also an unnerving note that outlined details of several unsolved murders that the writer claimed only he and the police knew. The note claimed that this information would prove he was the killer.

As it turned out, the cryptogram that the Zodiac enclosed was one-third of a three-part message—he mailed the other parts to the *San Francisco Examiner* and the *San Francisco Chronicle*. He demanded that the cryptogram—later to be referred to as the "Z-408" because it contained 408 characters—be published on the front page of the newspaper; otherwise, his note threatened he would go on a killing rampage "…until I have killed over a dozen people."[1]

Why would a person produce a cryptogram—a secret message—demand to have it published on the front page? Throughout the history of codes and ciphers, the

1. Dave Smith, *The Los Angeles Times*, "Zodiac the Killer," published in *The Tuscaloosa News*, October 17, 1969, p. 1.

correspondents tried to hide their messages from others as well as guarding the message's contents through the use of mathematical obscure cipher techniques should their dispatch be discovered. It made little sense why the Zodiac did this unless he had underlying psychological motives.

Though, from the perspective of our examination of codes and ciphers, the Zodiac code is a good case study. It helps demonstrates that anyone can create a cryptogram, but also how anyone can decrypt secret documents. It points to how easy mistakes can be made in the enciphering process, and with time and resources, all but a few crypto methods are unbreakable.

* * *

The Zodiac signed his note with a circle that was overlayed by a cross; the arms if the cross extending beyond the edge of the circle. The encrypted message comprised a combination of English and Greek letters, English letters that were reversed and upside-down; some astronomical and astrological symbols and symbols taken from ancient mythologies, such as the Egyptian Book of the Dead, works of the legendary continent of Mu, and the legends derived from primitive Pacific and Asian societies, as well as Native American rock paintings.

The Zodiac wrote in one of his notes, "…that when they do crack it, they will have me." The message was deciphered by Donald and Bettye Harden of Salinas, California. Mr Harden was a high school history and economics teacher, and with the help of his wife, they "cracked" the message on August 8, 1969. The Z-408 resembled many common ciphers and required the Harden's to deduce which symbols represented the alphabet letters (see Figure 1). The message, which contained spelling mistakes, was ungrammatical and even had coding errors, read:

> I like killing people because it is so much fun it is more fun than killing wild game in the forrest because man is the most dangeroue anamal of all to kill something gives me the most thrilling experence it is even better than getting your rocks off with a girl the best part of it is thae when I die I will be reborn in paradice and the I have killed will become my slaves I will not give you my name because you will try to sloi down or atop my collectiog of slaves for my afterlife ebeorietemethhpiti[2]

A psychiatrist from the Napa State Hospital found that the meaning of the Zodiac's signature symbol is likely to have represented the earth or perhaps the four elements that ancients thought comprised the world—earth, air, fire, and water.

A psychiatrist from the California Medical Facility in Vacaville analyzed the Zodiac's writings and concluded he was "...someone you would expect to be brooding and isolated. He is a guy who broods about cutoff feelings, about being cut off from his fellow man ... comparing the thrill of killing to the satisfaction of sex is usually an cxprcssion of inadcquacy."[3] Thcsc conclusions would explain the bizarre nature of the Zodiac's crimes and the likelihood he was criminally insane.

The Zodiac commented to the media that he was amused that his 408-cipher was cracked. Therefore, in his second encrypted message—referred to as the "Z-340"—he added more complexity; he divided it into three parts and included

---

2. This is the true decrypted message of the Zodiac's 408-symbol cipher, including the original's spelling and grammar errors (Robert Graysmith, *Zodiac* (New York: Berkley Books, 1976), pp. 54–55). At the time of this writing, the meaning, if any, of the last four words (ebeo riet emeth hpiti)—remains unsolved.

3. Dave Smith, *The Los Angeles Times*, "Zodiac the Killer," published in *The Tuscaloosa News*, October 17, 1969, p. 2.

several transpositions (see Figure 2). But it was an inadvertent enciphering error that introduced the real complication.

Figure 1—Copy of the Z-408 cipher.

The Z-408 cipher message was sent to the *San Francisco Chronicle* in 1969. This new cipher message took substantially longer to decipher; 51 years! It was solved in December 2020 by three amateur codebreakers: a Virginia-based software developer named David Oranchak; a

Belgian computer programmer—Jarl Van Eycke; and Australian mathematician—Sam Blake.

The Z-480 differed from the Z-340 in that the former's word alignment ran diagonally down the page, but occasionally it moved over one column. "It's a complicated bit of code creation," Mr Oranchak said, "but a basic scheme for it can be found in at least one U.S. Army code manual from the 1950s."[4] The deciphered message read:

Figure 2—Copy of the Z-340 cipher.

---

4. Daivid Oranchak quoted in Kevin Fagan, "Zodiac '340 Cipher' cracked by code experts 51 years after it was sent to the S.F. Chronicle," *San Francisco Chronicle*, December 20, 2020.

I hope you are having lots of fun in trying to catch me
That wasnt me on the tv show
Which brings up a point about me
I am not afraid of the gas chamber
Because it will send me to paradice all the sooner
Because i now have enough slaves to work for me
Where everyone else has nothing when they reach paradice
So they are afraid of death
I am not afraid because i know that my new life is
Life will be an easy one in paradice death

Arguably, the Zodiac killer is one of the more infamous unsolved murder cases not only in America but in the world. Nonetheless, his use of cryptography to communicate with the police and public makes this case of particular interest to this book.

Typically, cryptological writings prevent a third party from reading messages sent from one person to another. The intent of encrypting a message is to make it unreadable. Or unreadable long enough to perpetrate some action, thus making subsequent knowledge of the contents ineffective.

But the Zodiac's messages were not intended to do that because he communicated in plaintext and only sent two encrypted messages along with two shorter lines of code—one 13 characters long (Figure 3) and the other 32 characters (Figure 4).[5] It is not unreasonable to interpret this as him taunting the police, making them look feeble in the public's eyes.

5. i.e., Z-408 (July 31, 1969), Z-340 (November 8, 1969), Z-13 April 20, 1970), and Z-32 (June 26, 1970).

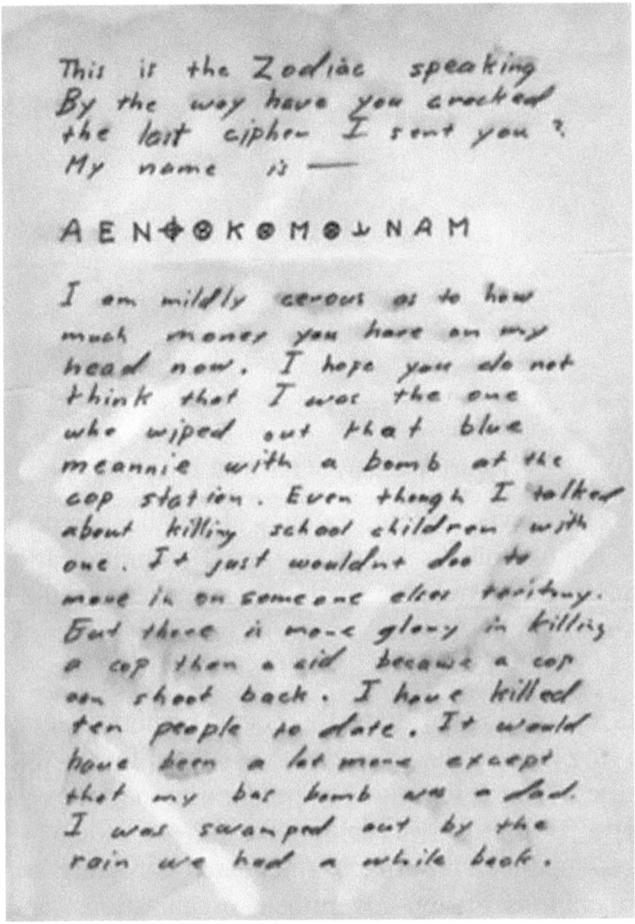

This is the Zodiac speaking
By the way have you cracked
the last cipher I sent you?
My name is —

A E N ⊕ ⊗ K ⊗ M ⊗ ⊥ N A M

I am mildly cerous as to how
much money you have on my
head now. I hope you do not
think that I was the one
who wiped out that blue
meannie with a bomb at the
cop station. Even though I talked
about killing school children with
one. It just wouldnt doo to
move in on someone elses teritory.
But there is more glory in killing
a cop than a cid because a cop
can shoot back. I have killed
ten people to date. It would
have been a lot more except
that my bus bomb was a dud.
I was swamped out by the
rain we had a while back.

Figure 3—Copy of the Z-13 cipher.

The fact that the Hardens—two amateur cryptogram enthusiasts—cracked the Z-408 code in a matter of days made the Zodiac devise a more complicated cipher. So, he did, but in doing so, he confused *himself* and introduced coding errors[6] that contributed to the long delay in cracking the cipher.

---

6. As pointed out in the U.S. Army's *Basic Cryptography* technical manual (TM 32-220, 1950), "…in some cipher systems

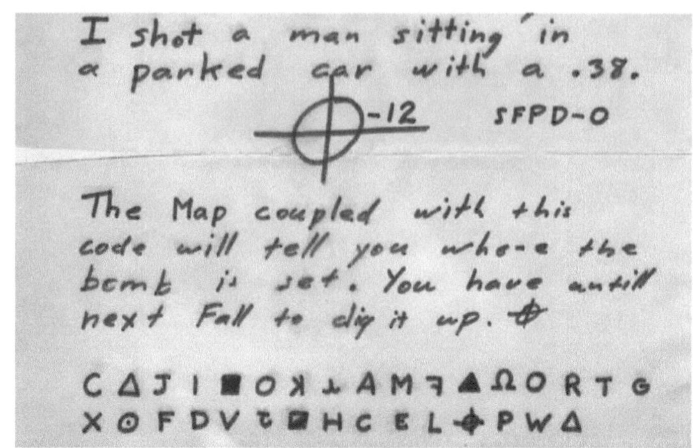

I shot a man sitting in a parked car with a .38.
-12 SFPD-0

The Map coupled with this code will tell you where the bomb is set. You have untill next Fall to dig it up.

Figure 4—Copy of the Z-32 cipher.

If the Zodiac was communicating with another person in the way cryptography is intended—secret communications—the other party would have noted the errors and would have drawn this to the Zodiac's attention, resulting in a corrected message being resent.

As a cryptological study, the Zodiac ciphers underscores why secret writing has provided the material for spy thrillers because it affords intriguing tropes for espionage novels and movies. Moreover, cryptography is an endeavor that protects nations and armies, safeguards corporate secrets, the operations of non-government organizations, and the private affairs of individuals.

So powerful are cryptological methods that countries have classified codes and ciphers as "weapons" and

---

a single error of a fundamental type, such as using the wrong key or the wrong 'setting,' may prevent the deciphering of the message." p. 74.

introduced legislative controls regarding their use and export.[7]

In this explanatory text on codes and ciphers, we will look at the world of codes and ciphers (Chapter 1), developing codes and ciphers (Chapter 2), breaking codes and ciphers (Chapter 3), the quest for an unbreakable cipher (Chapter 4), and devising an unbreakable cipher (Chapter 5).

---

7. And, a system of *key escrow* for digital communication equipment so that law enforcement agencies can access these technologies to conduct wiretaps and gather intelligence. Harold Abelson, et al., "Keys Under Doormats: Mandating Insecurity by Requiring Government Access to All Data and Communications, in *Journal of Cybersecurity*, Volume 1, Issue 1, September 2015, pp. 69–79.

— CHAPTER TWO —

WORLD OF CODES AND CIPHERS

"As we can scarcely imagine a time when there did not exist a necessity, or at least a desire, of transmitting information from one individual to another, in such manner as to elude general comprehension; so we may well suppose the practice of writing in cipher to be of great antiquity."[8]

Figure 5—Winston Churchill displaying "V for Victory" code.[9]

---

8. Edgar Allan Poe, "A Few Words on Secret Writing," *Graham's Magazine*, July 1841, Volume 19, p. 33.

9. Photograph compliments of the United Kingdom's Imperial War Museums (June 5, 1943).

During the Second World War, the British government broadcast the first four notes of Beethoven's Fifth Symphony, which was the letter V in Morse code—dot, dot, dot, dash—at the start of each British Broadcasting Corporation's (BBC) news transmission to Europe. This use of code was part of a broader "V for Victory" campaign used by the allies (see Figure 5). Although not secret in the sense of a coded message, it was a symbol that came to bind oppressed Europeans in their resistance to Nazi occupation.

The term *code* is used colloquially to refer to a multitude of methods for concealing the contents of messages. However, in cryptography, the code refers to replacing a message with a word or phrase that stands in for it. A coded message might be, "Vincent has a new dog." Any number of meanings could be attributed to this code phrase, such as, "The supplies will arrive tomorrow at the arranged time and place."

In contrast, a *cipher* is a method where a letter of the alphabet is replaced with another letter, or a number, a diagram, or a symbol, or a combination of all, as was the case of the Zodiac cipher in Chapter 1. Ciphers are characterized by a *key*. The cipher key is analogous to a building door key; only people with the key can access the contents on the other side of the door. A cipher key allows those who hold it to access the contents of the message.

An example of a simple cipher message might be 8, 5, 12, 12, 15. To decipher it, a cipher clerk would apply the key and unlock the message. In this case, the key is A = 1, so following the sequence dictated by the key (i.e., B = 2, C = 3. . .), the message reads, "Hello."

Where the key is unknown, as in the case of trying to crack an oppositions secret messages, this process is termed *cryptanalysis*. Cryptanalysis is both a science and an art. This is because it uses mathematical analysis as well as an understanding of the opposition—how they operate, their history of cryptographic algorithms, and their state of technical development, to help assess what might be the most fruitful way to approach solving the problem.

Traditionally, cryptanalysis was employed to crack secret military, spy, and diplomatic communications. But, it is also used to break the codes and ciphers of corporations involved in many forms of research and development, the operations of non-government organizations, and influential private individuals (e.g., people who are politically significant and leaders in innovation, commerce, and industry).

The military has a long history of using secret writing. One notable example from antiquity is the use of a *scytale*[10] by the Greek and Spartan armies (see Figure 6).

Figure 6—An example of the ancient Greek and Spartan scytale.[11]

A scytale is a wooden cylinder, like a baton. When a military campaign was launch, the emperor would give the general a scytale, and he would keep an identical. When the parties needed to communicate, they would wind a narrow

---

10. Also spelled *skytale*. John Laffin, *Codes and Ciphers: Secret Writing Through the Ages* (London: Abelard-Schuman, 1964), 18.

11. Photograph compliments of Wikipedia.

strip of papyrus, leather, or parchment around the scytale, and the message would be written longitudinally.[12]

When the parchment was unrolled, the message was unintelligible to anyone who might intercept the messenger on his way to the other party. But, when he safely reached his destination, the receiving party would wind the strip around their scytale to reveal the message.

This cipher method is a form of transpositional encipherment, which will be discussed in Chapter Three. Rather than using a theoretic key, it uses a physical key (i.e., the scytale) to achieve the displacement of the alphabetic characters.

It could be argued that the scytale was used as an aid to facilitate the enciphering and deciphering of messages rather than using a notional key. In support of this proposition, one need only look to the "rail fence cipher," which is the theoretical twin of the scytale; the latter using a device, the former a notional key.

As was seen when the Zodiac enciphered his messages, cipher clerks can also inadvertently introduce errors when using a key. However, using a physical device to assist the process reduces this risk of mistakes. Hence, over the centuries, several cipher devices were developed, many in the form of rotating cylinders and disks, or look-up tables and grids.[13]

The most well-known device was the Nazi Enigma machine. Several books have been written about this device and how the Polish *Biuro Szyfrów* (cipher bureau) was able

---

12. David Kahn, *The Code-Breakers* (New York: Macmillan Publishing, 1967), p. 82.

13. Such as the Wheatstone cryptograph and Plett's Cipher machine. Herbert O. Yardley, *The American Black Chamber* (Indianapolis: Bobbs-Merrill, 1931), p. 240.

to analyze the workings of the Enigma and pass these revelations on to allied intelligence.

Figure 7—Close-up of an Enigma machine's cipher wheels.[14]

Enigma was an electromechanical rotor device that allowed the operator to scramble the alphabet into a pseudorandom[15] (i.e., polyalphabetic) substitution cipher (see Figure 7). Mathematical calculations show that the Enigma system could result in more possible cypher combinations than the Allies could decrypt.[16]

---

14. Photograph compliments of Ted Coles.

15. *Random* (adj.), "Of or designating a phenomena the same outcome or consequence every time it occurs under identical circumstances (p. 1,080)." *Pseudo-* (adj.) "False… (p.1,055)" William Morris, editor, *The American Heritage Dictionary of the English Language* (Boston: Houghton Mifflin, 1971).

16. Josef Garlinski, *The Enigma War* (New York: Charles Scribner's Sons, 1980), p. 23.

> Each of the three rotors contained twenty-six letters or numbers. Since each rotor had twenty-six positions, the number of possible encoding positions increased to 17,256 (26 x 26 x 26). Since the Enigma came with six total rotors, that increased the total number of encoding positions to 105,456 (6 x 17 x 576). The reflector increased the number of encoding positions to over 7,000,000,000,000. The plug board could be set in over 100,000,000,000 ways. The final number of encoding positions for a normal Enigma machine equipped with three rotors, a reflector, and plug board was 5 x 10^{87}.[17]

"This was clearly sufficient to eliminate hand decryption as a possibility. The Allies would need to use machines of their own to decrypt the Enigma."[18]

The Enigma machine was the result of a long line of cryptosystems that date back to the first use of the written word.[19] It could be said that the invention of the alphabet and hence the written word was a form of code. Only those with the knowledge of the code—i.e., those who were literate—were able the encipher and decipher the message.

Take as an example Morse code. For those who have not studied the code, the dots-and-dashes are just a string of nonsense sounds. But to a Morse operator, the dits-and-dahs are heard as letters, just as clearly as the spoken word.

---

17. Walter A. Gagajewski, *British Cryptographic Efforts in World War II: The Struggle Against the German Enigma Machine*, Master's Thesis (Dominguez Hills: California State University, 1999), pp. 26–27.

18. John F. Dooley, *History of Cryptography and Cryptanalysis: Codes, Ciphers and Their Algorithms* (Cham, Switzerland: Springer International, 2018), pp. 151–152.

19. Michael E. Marotta, *Codes and Ciphers: All About Unbreakable Codes and How to Use Them* (Manson, MI: Loompanics Unlimited, 1979), p. 11.

As literacy spread, secret written communication methods evolved. It is estimated that the English vocabulary is over one million words. Nevertheless, it has been claimed that around ninety percent of all conversations can be understood if a person knows just three thousand of the most common words. This means that if a party wanted to hide the meaning of their message, they could use words that are not in common usage. By way of example, the medical terms used by surgeons to describe a procedure would baffle many people, even many university graduates.[20]

Once most of the population became literate, devices such as the scytale were needed to help hide the words, as was done by the Histiaios, the ruler of Miletus in the late-sixth-century B.C. While he was detained in the ancient city Persian city of Susa, Histiaios needed to send a message to his son-in-law, Aristagoras, in the Greek city of Miletus, instructing him to conduct an insurrection that Histiaios hoped would result in his release.

Because the roads to Miletus were guarded, any messenger that travelled them would be searched. Histiaios therefore devised a method of obfuscation known as *steganography*.

> For Histiaios, desiring to signify to Aristagoras that he should revolt, was not able to do it safely in any other way, because the roads were guarded, but shaved off the hair of the most faithful of his slaves, and having marked his head by pricking it [i.e., tattooing], waited till the hair had grown again; and as soon as it was grown, he sent him away to Miletos, giving him no other charge but this, namely that when he should have arrived at Miletos he

---

20. See footnote 64 regarding the use of special dictionaries in brute-force attacks.

> should bid Aristagoras shave his hair and look at his head: and the marks, as I have said before, signified revolt.[21]

Steganography is the method of concealing information, that is, hiding the presence of information. This can be done either physically—as in Histiaios' case—or digitally, such as hiding information in an image file, an audio file, or a word-processing file.[22] Steganography is concerned with hiding the presence of information, while cryptography deals with safeguarding information from unauthorized access.

If executed well, steganographic methods should completely hide the message, making the technique useful for situations where message discovery would pose a hazard for the bearer. An example that immediately presents is where an operative is crossing into or out of a prohibited area—a message could be encased in a capsule, swallowed, and, when passed, read.

On the surface, it might appear that steganography and encoding are conducted independently, but they can be combined to gain a higher level of information security. As we will discuss in Chapter Four, there are limitations to the effectiveness of cryptography, and there are limitations to steganography.

The first limitation is making the hidden information difficult to discover (i.e., the technique). The second is assessing how much information can be hidden before the volume makes it obvious.

---

21. "The Fifth Book of The Histories, Called Terpsichor," *The History of Herodotus*, translated by G.C. Macaulay (London: MacMillan and Co., 1890), paragraph 35.

22. Finn Brunton and Helen Nissenbaum, *Obfuscation: A User's Guide for Privacy and Protest* (Cambridge, MA: MIT Press, 2015).

Some techniques used in physical steganography included invisible ink, microdots, and null ciphers.[23] Most digital steganography techniques involved multimedia files, but also text files. For instance, a word processing document is an excellent place to bury a message. This could be done by typing the message, then changing the text color to white, which makes the message invisible against the document's white background. "Cover" text is then added—e.g., something innocuous like several of Shakespeare's sonnets—to divert attention, and the document is saved with a bland name.

Cryptography has a long history of being used by militaries, but probably the notable association is with intelligence agencies. Spy novels often feature the use of cryptograms or steganographic techniques.[24] Codes and ciphers are used as a counterintelligence system to guard the information. It is also used in offensive deception operations.

Referred to as *counterespionage*, one aspect of this intelligence function is to neutralize the opposition, "… not just by destruction, but also by paralysis—disruption."[25] Disruptions are actions designed to frustrate hostile intelligence operations. As an example, take the case of

---

23. Although invisible ink and microdots are well known, null ciphers are less recognized. A null cipher is a method concealing a plaintext message amongst a larger amount of non-cipher material. It is a form of hiding in plain sight and hence sometimes termed the *concealment cipher*.

24. e.g., Robert Harris, *Enigma* (New York: Random House, 1995).

25. Hank Prunckun, *Counterintelligence Theory and Practice, Second Edition* (Lanham, MD: Rowman & Littlefield, 2019), p. 223.

numbers stations.[26] These are shortwave[27] (SW) radio stations that operate as one-way voice links between "handlers" and their field agents, who are often working in denied areas.

The transmissions contain a long string of numbers, usually sent in groups of five numbers. These numbers are broadcast at a pre-arranged date and time. Using a simple AM-FM-SW receiver such as the one shown in Figure 8, the theory is that the field agent, or deep cover operative, receives the numbers and deciphers the message. If the field operative maintains sound operational tradecraft, especially having a plausible reason to have a shortwave receiver, and is not subject to electronic surveillance, then a shortwave *one-way voice link* is a secure communications system.

Nevertheless, opposition intelligence services still monitor transmissions and record the enciphered messages in the hope of either cracking the message or obtaining the key sometime in the future so archived messages can be read.[28] Yet, there are faux messages transmitted in the hope they will mislead the opposition and trick them into wasting computer and cryptanalytic resources trying to decipher the transmissions. If enough of these fake messages are sent, some level of paralysis can be achieved.

It is not unreasonable to assume that diplomatic missions of all countries encrypt some portion of their incoming and outgoing message traffic. It has been estimated that as much as ninety-percent of diplomatic electronic

---

26. Also termed *clandestine one-way broadcasts*. John T. Corley, *Clandestine Communications Systems*, Master's Thesis (Monterey, CA: U.S. Naval Postgraduate School, 2001), p. 37.

27. The shortwave bands are between 3 MHz and 30 MHz.

28. Winston Smith, *Covert Techniques for Intercepting Communications* (Commack, NY: CRB Research, 1981), p. 10.

correspondence is encrypted. The ten-percent that is transmitted "in the clear" is done so,

> ...in an attempt to reassure everybody that nations of the world are making at least some minimal attempt to smooth over the ripples and wrinkles that constantly pop up in the day-to-day ideological, political, commercial, and philosophical relations with one another.[29]

These plaintext messages involve such public relations topics as "...treaty signing, VIP visits, international conferences, relief efforts, trade arrangements, scientific cooperation, sports competitions, the protection of endangered wildlife species, and the overall 'hand-shakes and smiles' aspects of international diplomacy."[30]

Figure 8—Sangean ATS 909X Portable World Band Receiver.[31] Courtesy of Sangean Electronics, Inc.

---

29. Tom Kneitel, *Guide to Embassy and Espionage Communications* (Commack, NY: CRB Research, 1986), p. 5.

30. Tom Kneitel, *Guide to Embassy and Espionage Communications*, p. 5.

31. This is one example of many small portable receivers that are popular with travelers because they operate on batteries and receives AM, FM, longwave, and shortwave broadcasts. Such a device is unlikely to attract unwanted attention when used for receiving numbers stations.

The encrypted traffic is almost certain to feature more sophisticated methods than the simple substitution or transpositional ciphers of the early twentieth-century. This is because nations in the Five Eyes intelligence alliance (FVEY)[32] as well as the Russian Federation and Communist China spend very large amounts of time and resources to crack government and diplomatic messages.

Yet, smaller nations without an intelligence or military alliance with a major world power, or is a member of, say, NATO, would have to purchase equipment from a commercial supplier. Although this equipment is excellent for many common business communications, the cryptograms these systems generate would be no match for the analytical prowess of the United States, Russia, or China.[33]

For businesses, private sector consultants, and non-government organizations, commercially available crypto devices are likely to be more than adequate. Many of the devices on the market have been approved by various government authorities for message traffic up to *top secret*.[34]

Cryptosystems have even been used by individuals who are concerned about over-zealous government authority and what they see as the growth of state power. There are several types of groups who have such a political outlook—both on the right and left of politics—that view autonomy, individualism, political freedom, and other manifestations of liberty as their philosophical principles.

---

32. The Five Eyes countries comprise Australia, Canada, New Zealand, the United Kingdom, and the United States.

33. And, perhaps, nations such as Israel.

34. *Top secret* is usually reserved for the most sensitive information transmitted by a nation. See footnote 35.

Given this type of position, it follows that some of these ideologues would see fit to communicate with each other using codes. Unlike the governments they profess to distrust, it is unlikely that these people have the financial resources to buy the technology that would allow them to converse securely—arguably only a handful of world powers have that capability. Even if libertarian-type ideologists could purchase next best level of security—commercial-grade encryption devices—these are not *totally* secure from decryption by some governments.

Although these groups' systems might be able to baffle a local police or sheriff's department, any attempt to communicate secretly would be analogous to using a cipher found in the puzzles and cryptogram section of a newspaper or magazine. The reliance on rudimentary types of encryption suggests that users need a more developed understanding of how codes and ciphers are created. Understanding how these techniques are developed assists in determining which method is needed to protect different levels of communications.[35] Protection levels, of course, is based on an assessment of vulnerability.[36]

---

35. Information is classified according to the potential harm unauthorized disclosure could cause. Various nations (as well as corporations) have definitions that suit their needs; but as an example, levels may resemble these: Unclassified; Official; Restricted; Confidential; Secret; and Top Secret. A classification of *Secret* would be defined as information that could cause "serious damage" to national security.

36. A vulnerability assessment evaluates the relationship between three variables—target attractiveness, ease of attack, and impact. See, Hank Prunckun, *Methods of Inquiry for Intelligence Analysis, Third Edition* (Lanham, MD: Rowman & Littlefield, 2019), pp. 193–199.

— CHAPTER THREE —

DEVELOPING CODES AND CIPHERS

Ciphers are classified as either being a substitution cipher, a transpositional cipher, or a concealment cipher.[37] Substitution is the process of replacing plaintext letters with other symbols. Transposition involves taking the plaintext letters out of their natural order and rearranging them according to a *key*—an agreed pattern established by the sender and recipient. A concealment cipher is a form of encryption that hides the plaintext amongst non-cipher material. Hence, it is also called a *null cipher*. Some cryptographers, therefore, consider it a form of steganography.

A code is a particular type of cipher. It is a system that replaced words, phrases, or sentences with briefer representations—i.e., code words/numbers/phrases. This could be done simply to save time, or as with a secret message, to hide the true meaning. An example of a code being used to save time is the Q-code used by amateur radio operators. QSL means "all understood"; QRP means "operating under reduced transmission power"; QRN means "signals are being affected by atmospheric static"; and so on. There are several dozen Q codes in use.

Semaphore is a visual code like the audio code produced by the telegraph—Morse code. It was developed by the French engineer Claude Chappe in 1794 as a means of transmitting messages between Paris and the Lille during the French Revolution.

---

37. Helen Fouché Gaines, *Cryptanalysis: A Study of Ciphers and their Solutions* (New York: Dover Publications, 1956), p. 1.

The original system comprised a series of towers that, at the top, supported a two-armed *semaphore*.[38] Each arm was able to display seven positions, making for a total of forty-nine combinations. Each position was designated a letter of the alphabet, a number, or a symbol.

To transmit a message, the first tower operator would signal the second tower several miles away using the arms. The second semaphore operator would read the message using a telescope and resend it to the third tower, and on it would go until it reached its destination. The reply would occur the same way, only in reverse.

Semaphore has been used to communicate between ships because it was a convenient means of sending messages while maintaining radio silence.[39] The signaler can use small handheld flags, paddles, disks rods or even their hands (see Figure 9). If illuminated, such devices can be used to send a message at night.

Morse code has been used since 1844 when Samuel Morse sent the first coded message via telegraph between Washington to Baltimore. It is the audio equivalent to semaphore, which is visual. Originally used in telegraphy, especially with railroads in America and Canada. The alphabet, numerals, and symbols were converted to dits-and-dahs in the same manner that these characters were converted to semaphore positions.

---

38. This is the term Chappe coined based on the Greek word for "bearing a sign."

39. *Radio silence* refers to cession of transmission where such signals could reveal troop positions (i.e., using direction finding equipment), troop strengths, and/or troop movements. Although a security measure, the fact that an adversary has gone radio silent can be interpreted by radio traffic analysts as the prelude to a noteworthy event, such as a surprise attack.

Although not a secret code, Morse code does offer some level of security in *public* use, as does using words and phrases that are not in common circulation, or speaking in a foreign language.[40] But acquiring a more extensive vocabulary, learning a new language, or studying Morse code will lay bare any such disguised message.

Figure 9—Quartermaster Seaman Ryan Ruona using semaphore flags during a replenishment at sea aboard the aircraft carrier USS Theodore Roosevelt in the Persian Gulf, 2005. Photo courtesy U.S. Navy by Photographer's Mate Airman Javier Capella.

Nevertheless, Morse code and semaphore are examples of *substitution ciphers*. As the name suggests, a letter, number or symbol is substituted with another character or sound. One of the first uses of substitution cipher was the *Caesar*

---

40. Recall is the method that the U.S. military used in the Pacific Theater during World War II. To send messages that could not be understood by the enemy, Navajo code-speakers were employed. A similar method was used by the Libyans by using the Berber language. But unlike the Navajo language, the Berber language was known to intercept stations, so the tactic failed. Tom Kneitel, *Guide to Embassy and Espionage Communications* (Commack, NY: CRB Research, 1986), p. 25.

cipher. The figure below shows diagrammatically how the cipher works. In this example, the letter A is shifted three places along in the alphabet to align with the letter X. B is moved to align with the letter Y, and C is realigned with Z. As the figure shows, the other letters follow suit.

Plaintext Alphabet	a	b	c	d	e	f	g	h	i	j	k	l	m	n	o	p	q	r	s	t	u	v	w	x	y	z
Ciphertext Alphabet	D	E	F	G	H	I	J	K	L	M	N	O	P	Q	R	S	T	U	V	W	X	Y	Z	A	B	C

Figure 10—Caesar cipher with a shift of three positions. The plaintext is shown at the top, and the ciphertext is displayed at the bottom. Variants include shifts in the opposite direction, or an increase or decrease in the places moved.

Figure 10 shows diagrammatically how the cipher works. In this instance, the letter A is shifted three places along in the alphabet to align with the letter D. B is moved to align with the letter E, and C is realigned with F. The other letters follow suit.[41]

The use of a substitution cipher transforms a plaintext message into a collection of characters that are incomprehensible (and hence, secret). To decipher the message, it takes more than learning, say, Morse code; it takes some understanding of how ciphers are created.

Substitution ciphers are simple in concept. A basic cipher is certainly an adequate method for young people who want to send notes in the school classroom when they are bored. The risk of their teacher spending time deciphering a note is low—they usually are overworked. If, however, if a diplomat at a foreign post was to use the same system, adversaries would have no trouble deciphering it.

---

41. Although the method of substitution is shown in a two-line format, this can be accomplished in the field by using two concentric disks—one larger than the other, with the smaller one mounted on top. In the way, the smaller disk can be *rotated* (rather than *shifted*) to align with the correct key.

The methods for deciphering secret messages will be discussed in "Chapter Four."

Another type of cipher is transpositional. These ciphers differ from substitution ciphers in that the former jumbles the letters of the plaintext message rather than replacing the plaintext letters with other letters as is done in the latter.[42]

To demonstrate this technique, take the plaintext message of "We need to meet." Using a simple two-column transpositional method, a cipher clerk will arrange the letters of the plaintext in two columns eliminating spaces and punctuation marks as shown below.

WE
NE
ED
TO
ME
ET

This jumbled—transpositional—arrangement facilitates the encipherment by reading down the first column and then down the second to arrive at this ciphertext: "WNETMEEEEDOET." Although at first glance, the text looks indecipherable, a person skilled in the boardgame Scrabble® could work out the message.

So, if substitutional ciphers replace a ciphertext symbol with a plaintext symbol, then a transpositional cipher uses permutation of plaintext letters in a rearrangement without altering the symbols' true representation. This is applicable to not only information transmitted in the physical world—

---

42. Norman Bruce, *Secret Warfare: The Battle of Codes and Ciphers* (Melbourne: Wren, 1973), p. 14.

including radio transmissions—but also the bits and bytes of electronic information.

To add complexity to the method, variants can be introduced using keywords that add columns and rearrange the columns' reading order. Or, the order can oscillate like the *rail fence* method.[43] This technique uses descending-and-ascending diagonal columns to encipher messages. For instance, the message "spring follows winter" (as shown in Figure 11) would result in the following ciphertext—"SNOWIRPIGFLOSWNER L T." Still, a relative insecure message.

S	-	-	-	N	-	-	-	O	-	-	-	W	-	-	-	I	-	-	-	R
-	P	-	I	-	G	-	F	-	L	-	O	-	S	-	W	-	N	-	E	-
-	-	R	-	-	-		-	-	-	L	-	-	-		-	-	-	T	-	-

Figure 11—Example of the rail fence variant of a transpositional cipher.

The next level of complexity is to combine substitutional and transpositional ciphers, so the plaintext message undergoes a double encipherment process. Even this is likely to prove insecure to a well-resourced opponent.

Another method of encryption uses *codes*. Coding is also a method of substitution, but rather than rearranging the message's word-lettering pattern, it replaces a word, a phrase, or even a whole sentence with one or more words. For instance, the message to meet at noon at the village's café could be coded as "rose peddles."

Cryptographers refer to this method as a monoalphabetic cipher because it uses fixed substitution for the whole

---

43. It has been reported that this modification was used up to the U.S. Civil War (Michael E. Marotta, *Codes and Ciphers: All About Unbreakable Codes and How to Use Them*, p. 8).

message. While simple substitutional ciphers can be relatively easy to solve, coded message are not.

Decrypting a code relies on receiving many messages, long messages, or both. In this way, techniques such as repetition can help define the code's structure. Or, breakthroughs involving a few code groups can also help predict other code groups. This method is referred to as a "window index." "Every time a word or phrase was broken out, it was indexed to everywhere else it appeared in the matched traffic."[44]

Moreover, after-event analysis of preceding coded messages may be able to yield meaning. Nonetheless, this is a time-consuming, intuitive, task that has a low probability of success.

Regardless, as Benjamin Franklin is attributed with saying, "Three may keep a secret if two of them are dead,"[45] has significance to codes. That is because codes require codebooks—large bound texts that contain both the plaintext and its code word/group. These codebooks need to be distributed to all users, and it is here that Franklin's saying has meaning. The more people who have a cookbook, the greater the risk that one of these texts may fall prey to the opposition's attempts to acquire them.[46]

In this regard, defensive counterintelligence—comprising physical security, personnel security, information security, and communications security—

---

44. Peter Wright with Paul Greengrass, *Spy Catcher: The Candid Autobiography of a Senior Intelligence Officer* (Richmond, Victoria: William Heinemann Australia, 1987), p. 181.

45. Gerard Gawalt, *In His Own Words Library Exhibition Celebrates Tercentenary of Benjamin Franklin's Birth*, (Washington, DC: Library of Congress, 2006).

46. e.g., robbery, burglary, bribery, or capture.

becomes vital.[47] If, however, a codebook is compromised, then all codebooks in circulation need to be replaced. Before the age of ubiquitous computing, this was an arduous pencil-and-paper task. As can be imagined, the job is the equivalent of writing a new dictionary.[48] Then, there is the problematic work of distributing the new codebooks, including all the security risks that accompanied this process.

Creating a new codebook has been made more manageable with computers, but it is still not an easy task. Having said that, codebooks are retired at irregular intervals to frustrate hostility cryptanalysts' efforts to crack the code and to limit the potential damage of an undetected leaked/embezzled codebook. So, the work in cryptographic units is a persistent job.

This was the case in January 1917. The German Foreign Minister, Arthur Zimmermann, send a telegram to the President of Mexico suggesting the two countries enter a military alliance against the United States. In exchange for Mexico's support in the "Great War," Germany pledged to assist Mexico in reclaiming Texas, Arizona, and New Mexico. However, the British intercepted the coded telegram, were able to decipher the message, and then briefed the Americans.

---

47. See Hank Prunckun, *Counterintelligence Theory and Practice, Second Edition* (Lanham, MD: Rowman & Littlefield, 2019), pp. 88–172.

48. Even with computers, the production of a new code book is time consuming and laborious because the contents need to be checked by hand to ensure accuracy, eliminate repetition, or close association with events, etc. As for creating codebooks in advance ready for distribution, there is a risk that these could be leaked to the opposition or hacked by adversaries. So, a "print-on-demand" system would be a counterintelligence solution.

Known as the "Zimmermann Telegram," it was coded using number groups instead of code words. The declassified four- and five-number code groups is shown in the figure below. Notoriously difficult to crack, the British success was because one of their overseas agents was working in a Brussel's radio station. The station was responsible for transmitting German messages, and an Austrian-born British agent—Alexander Szek—had access to the codebook. He memorized a few words each day and passed them onto his handlers.[49]

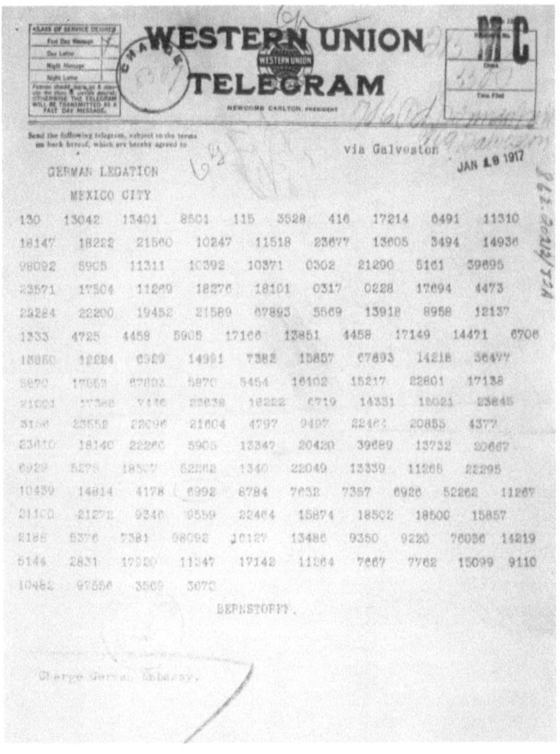

WESTERN UNION
TELEGRAM
NEWCOMB CARLTON, PRESIDENT

Send the following telegram, subject to the terms on back hereof, which are hereby agreed to

via Galveston

JAN 19 1917

GERMAN LEGATION

MEXICO CITY

130 13042 13401 8501 115 3528 416 17214 6491 11310
18147 18222 21560 10247 11518 23677 13605 3494 14936
98092 5905 11311 10392 10371 0302 21290 5161 39695
23571 17504 11269 18276 18101 0317 0228 17694 4473
22284 22200 19452 21589 67893 5569 13918 8958 12137
1333 4725 4458 5905 17166 13851 4458 17149 14471 6706
13850 12224 6929 14991 7382 15857 67893 14218 36477
5870 17553 67893 5870 5454 16102 15217 22801 17138
21001 17388 7446 23638 18222 6719 14331 15021 23845
3156 23552 22096 21604 4797 9497 22464 20855 4377
23610 18140 22260 5905 13347 20420 39689 13732 20667
6929 5275 18507 52262 1340 22049 13339 11265 22295
10439 14814 4178 6992 8784 7632 7357 6926 52262 11267
21100 21272 9346 9559 22464 15874 18502 18500 15857
2188 5376 7381 98092 16127 13486 9350 9220 76036 14219
5144 2831 17920 11347 17142 11264 7667 7762 15099 9110
10482 97556 3569 3670

BERNSTORFF.

Charge German Embassy.

Figure 12—The Historic 19 January 1917 "Zimmermann Telegram," Courtesy of the U.S. Department of State.

49. John Laffin, *Codes and Ciphers: Secret Writing Through the Ages* (London: Abelard-Schuman, 1964), p. 95.

Once the contents of the Zimmermann telegram were made public, the Germans started hunting for the mole. Before their counterintelligence operatives closed in on Szek, he was exfiltrated from Belgium to France, but when that happened, the Germans no doubt realized who leaked the code and changed it.

Nonetheless, there were examples where the Germans were unaware their codebooks had been seized. This happened during the Battle of Jutland where the British obtained a copy of the German navy's codebook from the Russian light cruiser SMS *Magdeburg*, which the Russian's boarded after the *Magdeburg* ran aground in Russian territorial waters.[50] Such feats underscore the need to change codes and ciphers at intervals that cannot be predicted.

---

50. Johnathan Sutherland and Diana Canwell, *The Battle of Jutland* (South Yorkshire: Pen & Sword Maritime, 2007).

— CHAPTER FOUR —

BREAKING CODES AND CIPHERS

It follows that if a mathematical system can convert a plaintext message into a cryptogram, then the reciprocal is true—the cryptogram can be analyzed to reveal the secrets its hides. This outlook was captured by Herbert O. Yardley when he is reported to have said, "I always assume that what is in the power of one man to do is in the power of another."[51]

Many methods can be used to solve cryptograms—some relying only on intuition or guesswork, others employing methodological pencil-and-paper doggedness, while others require computer resources. Collectively, these art-and-science processes are known as *cryptanalysis*. Nonetheless, the objective is more than solving a single cryptogram but recovering the cipher key so that it can be applied to all encrypted traffic.

When cryptologists develop a cipher, they do so with the understanding that their adversaries will attempt to intercept the cryptograms and decipher them. Therefore, cryptographers try to invent new or novel ways to make it as difficult as possible for their opponents.

Radio engineers also try to devise methods for transmitting messages that limit the opposition's ability to intercept the encrypted signals.[52] Regardless, it should be

---

51. James Bamford, *The Puzzle Palace* (Boston: Houghton Mifflin Company, 1982), p. 7.

52. Take for instance, a technique used by the Germans during the Great War that is still used today. British listening posts seemed to hear what they considered atmospheric static but was cryptographic traffic that was sped-up. The receiving station

noted that cryptographers try to make it more *challenging*. Unbreakable codes and ciphers will be discussed in "Chapter Five,"; still, other than a spy or agent acquiring a copy of the cipher key or codebook, some of the primary ways cryptograms are broken will be examined here.

While television programs show protagonists solving cryptological problems in the thirty- or sixty-minutes (the time in which the show airs). Regardless, the reality is that cryptograms take much longer and involve backtracking and restarting along with long periods of thinking and guessing. Breaking cryptograms takes not only an understanding of how codes and ciphers are made, but it requires an analyst to have a claim, patient personality. Above all, an analyst needs the same level of perseverance that a SWAT officer or special forces soldier has in order to keep going when all seems hopeless.

Like commandos who have special training in using tactical equipment, a cryptanalyst benefits from having specialist knowledge in language—English or whatever language that may be being decrypted. This is because some techniques require an understanding of how words are used, spelled, sentences punctuated, and the grammatical conventions of sentence construction.

Take, for instance, what is referred to as a language's redundancy. Redundancy implies that there are words that can be omitted, yet the meaning of the message can still be understood. For example, "added bonus" because a bonus is something added.

---

would replay the message at a slower speed. Until this was realized by the British, many messages flew through the ether undetected. This could be considered the precursor to "burst transmissions." See, Department of the Army, *Code-Burst Transmission Group, TM-11-5835-224-35* (Washington, DC: Department of the Army, 1969).

Other peculiarities of the English language included words that, in almost all cases, begin the same, such as words starting with "q" are regularly followed by the letter "u." There are exceptions, but from an analytic perspective, if an encrypted word is suspected of beginning with "q," then it is highly likely that the next letter is "u."

There are other idiosyncrasies that can aid decryption. For instance, there are some letter combinations that do not appear in *English* words, such as words starting with the letters "ng."[53]

Moreover, English depends on many words of four letters or less—like the indefinite articles "a," and "an"; the definitive article "the"; the conjunction "and"; the pronouns "I," "he," "she," and "it"; as well as the prepositions "at," "by," "for," "of," "in," "into," "to," and "with"; the pronoun/ determiner/ conjunction/ adverb "that," and the state of being verb "is"—that can account for one-quarter of the words in an average English plaintext message.[54]

Regardless of the complexity of a cipher, understanding grammar will help decryption. This is because sentences are constructed of a subject, verb, and object. Knowing the word pattern in which sentences are likely to take gives cryptanalysts another potential starting point when they search for solutions.

---

53. Though there are many foreign words, such as *ngoni*, which is a West African word for a guitar-like stringed instrument; *Ngai* is the transliteration of three Cantonese surname variants that is found predominantly in Hong Kong (*Ng* is also a Vietnamese surname); *ngaio* is a Mauri word for a New Zealand evergreen tree (and a girl's first name); and *ngwee* is the name of a Zambian monetary unit.

54. David Kahn, *The Code-Breakers* (New York: Macmillan Publishing, 1967).

One of the best-known techniques in cryptanalysis is the use of *frequency analysis*. This method, discovered in the 870s by the Muslim philosopher Abu al-Kindī,[55] is grounded in the theory of statistical inference. That is to say, in every language, there are letters that appear more often than others. In English, the letter "e" is the letter that is most frequently used, followed by the letters "t," "a," "o," "i," "n," "s," "h," and "r." The least used letters are "z," "q," "x," "j," and "k."[56]

Knowing this fact has allowed cryptanalysts to construct frequency tables for not only single letters, but also for double (diagram), and triple (trigram) letters. This is also done for the languages used by adversaries. They do this by taking a sample of plaintext messages of the sort that they anticipate they will encounter and count the characters making a tallying of each letter.[57]

If the letter distribution of the ciphertext is comparable to the frequency found in a sample of plaintext messages (e.g., taken from, say, a sample of news articles), then a cryptanalyst could conclude that it is a monoalphabetic cipher or a simple substitution cipher.

Recall from our discussion in the previous chapter of the Caesar cipher; substitution ciphers replace each letter of the plaintext with another letter. It is a one-to-one relationship—"A" is substituted with "B," "B" with "C,"

---

55. *Manuscript on Deciphering Cryptographic Messages.*

56. Derived from sample passages from newspapers and novels that totaled 100,362 alphabetic characters. Henry Beker and Fred Piper, *Cipher Systems: The Protection of Communications* (London: Northwood Books, 1982).

57. This could be done by using a reasonably large sample of decrypted messages.

"C" with "D," and so on.[58] In a polyalphabetic cipher, the substitution relationship is one-to-many (essentially, a series of Caesar ciphers), and hence, more complicated to decipher.

Its development is credited to the German monk Johannes Trithemius, the Italian cryptologist Giovan Battista Bellaso, and French diplomat Blaise de Vigenère during the 1500s. The process comprises a *tabula recta* and a key. Like other ciphers, the key is what determines which of the twenty-six alphabets contained in the table are used to decipher the secret message. A typical table is shown in Figure 13.

	A	B	C	D	E	F	G	H	I	J	K	L	M	N	O	P	Q	R	S	T	U	V	W	X	Y	Z
A	A	B	C	D	E	F	G	H	I	J	K	L	M	N	O	P	Q	R	S	T	U	V	W	X	Y	Z
B	B	C	D	E	F	G	H	I	J	K	L	M	N	O	P	Q	R	S	T	U	V	W	X	Y	Z	A
C	C	D	E	F	G	H	I	J	K	L	M	N	O	P	Q	R	S	T	U	V	W	X	Y	Z	A	B
D	D	E	F	G	H	I	J	K	L	M	N	O	P	Q	R	S	T	U	V	W	X	Y	Z	A	B	C
E	E	F	G	H	I	J	K	L	M	N	O	P	Q	R	S	T	U	V	W	X	Y	Z	A	B	C	D
F	F	G	H	I	J	K	L	M	N	O	P	Q	R	S	T	U	V	W	X	Y	Z	A	B	C	D	E
G	G	H	I	J	K	L	M	N	O	P	Q	R	S	T	U	V	W	X	Y	Z	A	B	C	D	E	F
H	H	I	J	K	L	M	N	O	P	Q	R	S	T	U	V	W	X	Y	Z	A	B	C	D	E	F	G
I	I	J	K	L	M	N	O	P	Q	R	S	T	U	V	W	X	Y	Z	A	B	C	D	E	F	G	H
J	J	K	L	M	N	O	P	Q	R	S	T	U	V	W	X	Y	Z	A	B	C	D	E	F	G	H	I
K	K	L	M	N	O	P	Q	R	S	T	U	V	W	X	Y	Z	A	B	C	D	E	F	G	H	I	J
L	L	M	N	O	P	Q	R	S	T	U	V	W	X	Y	Z	A	B	C	D	E	F	G	H	I	J	K
M	M	N	O	P	Q	R	S	T	U	V	W	X	Y	Z	A	B	C	D	E	F	G	H	I	J	K	L
N	N	O	P	Q	R	S	T	U	V	W	X	Y	Z	A	B	C	D	E	F	G	H	I	J	K	L	M
O	O	P	Q	R	S	T	U	V	W	X	Y	Z	A	B	C	D	E	F	G	H	I	J	K	L	M	N
P	P	Q	R	S	T	U	V	W	X	Y	Z	A	B	C	D	E	F	G	H	I	J	K	L	M	N	O
Q	Q	R	S	T	U	V	W	X	Y	Z	A	B	C	D	E	F	G	H	I	J	K	L	M	N	O	P
R	R	S	T	U	V	W	X	Y	Z	A	B	C	D	E	F	G	H	I	J	K	L	M	N	O	P	Q
S	S	T	U	V	W	X	Y	Z	A	B	C	D	E	F	G	H	I	J	K	L	M	N	O	P	Q	R
T	T	U	V	W	X	Y	Z	A	B	C	D	E	F	G	H	I	J	K	L	M	N	O	P	Q	R	S
U	U	V	W	X	Y	Z	A	B	C	D	E	F	G	H	I	J	K	L	M	N	O	P	Q	R	S	T
V	V	W	X	Y	Z	A	B	C	D	E	F	G	H	I	J	K	L	M	N	O	P	Q	R	S	T	U
W	W	X	Y	Z	A	B	C	D	E	F	G	H	I	J	K	L	M	N	O	P	Q	R	S	T	U	V
X	X	Y	Z	A	B	C	D	E	F	G	H	I	J	K	L	M	N	O	P	Q	R	S	T	U	V	W
Y	Y	Z	A	B	C	D	E	F	G	H	I	J	K	L	M	N	O	P	Q	R	S	T	U	V	W	X
Z	Z	A	B	C	D	E	F	G	H	I	J	K	L	M	N	O	P	Q	R	S	T	U	V	W	X	Y

Figure 13—An Example of a Trithemius Cipher Table

58. Or, instead of shifting the alphabet a given number of letters right or left, a cryptographer could arbitrarily allocate letters.

The polyalphabetic substitution cipher is the technique the Nazis used for their Enigma machine. The Enigma device repeatedly changed its electrical paths to rearrange the alphabet in a pseudorandom form.

But despite their complexity over simple substitution ciphers, these are still vulnerable. This is because polyalphabetic ciphers require a keyword to guide the encryption and decryption process. So, if the keyword is stolen, leaked, or guessed, then all messages based on it will be laid bare. The counter to this, the time-tested defence has been to change the keyword often, but at irregular intervals.[59]

Changing keywords is one of the first principles of cryptographic security, but there are many techniques that cryptographers use to confound an opposition's cryptanalysts. One method is to use a *polyphone*, as was the case with the Zodiac cipher we discussed in "Chapter One." This technique substitutes a plaintext letter with a symbol. For instance, the symbol # might replace the letters "w" *and* "v."

A variation is to use *homophonic substitution.* That is, different cipher symbols are used to represent a single plaintext letter. The symbols "◆" *and* "✡" might represent the single letter "u." This technique is used in the Vigenère cipher, where there is a many-to-one relationship.

Cryptographers also use the technique of employing *null symbols*, which represent no plaintext letters; their use just makes the cipher more obscure. For instance, the symbol ⮤ (along with others) might be inserted randomly in the ciphertext. The decrypting station would ignore the symbol, but the intercepting party would not know this and

---

59. *Not*, for example, *everyone* Monday at 0800 hours UTC, or any other scheduled day and time.

would add some level of complexity (i.e., time, effort, and resources/money) trying to unravel the key.

So far, we have discussed some general aspects of cryptanalytic "attacks".[60] Now it is time to explain these in more detail.

The cryptanalytic approach is a reverse-engineering enterprise. The techniques that comprise this method acknowledge Kerkhoff's Doctrine,[61] so, in the main, attacks on the adversary's crypto-system(s) aims to recover the key.

The current information about the system determines the method used.[62] Nevertheless, even if the key cannot be recovered, the fact that parties of interest are engaged in secret communication can provide enough information to facilitate analysis. This is termed *traffic analysis*—"who is talking to who," "how often do they communicate," and what is the order of their message traffic.[63] Based on these

---

60. The term *attack* is used to describe the process of finding a weakness in a code, cipher, or crypto-key.

61. Kerkhoff's Doctrine (or Principle) theorizes that a cryptosystem's security will remain secure even when all aspects of the system are known to the opposition, *except the key*. A version of this was put forward by cryptographer Claude Elwood Shannon, known as *Shannon's maxim*, as "The enemy knows the system." *Shannon's Maxim*, in H.C.A. van Tilborg and S. Jajodia (eds.), *Encyclopedia of Cryptography and Security* (Boston: Springer, 2011).

62. Other ways of attaching a cryptosystem are to use attack the implementation via a *side-channel attack* (i.e., interpretation of the hardware's technical emissions or data) or target to users through the use of deception or a pretext.

63. U.S. Department of the Army, *Fundamentals of Traffic Analysis (Radio-Telegraphy) TM 32-250*, reprinted by Aegean Park Press, Laguna Hills, CA.

data, inferences can be drawn about the parties' disposition and plans.

The information at hand will determine the technique used. For instance, if the analyst has little or no information about the cryptogram, then a *ciphertext-only attack* is usually performed.

The most common method to do this is by way of a *brute-force* search.[64] This is done by trying every possible key.[65] If some intelligible text is yielded, then closer examination may be able to fully solve the key.

Nonetheless, it has been estimated that, on average, half of all possible keys must be tried to successful crack a cipher using brute force.[66] The calculation supporting this assertion is thus: assume N is the total number of possible random keys. Then the probability that any one key could be the correct key is 1/N. If a cryptanalyst investigates half the potential number of key (i.e., N/2), her chance of finding the right key is (1/N) x (N/2), which equates to 1/2, or in percentage terms, 50%.

If the target of the attack is, say, a four-digit telephone passcode, then it can be accomplished quickly because there

---

64. *Brute force* refers to a technique that uses an automated process based on trial-and-error. The essence of the method is to "guess" the target's username, password or passphrase, credit-card number, or, in terms of a cipher, the cryptographic key. The process of guessing may take the form of a *dictionary attack*. That is, systemically trying every word in a dictionary. In doing so, many dictionaries can be tried: medical dictionaries, law dictionaries, engineering dictionaries, various slang dictionaries, dictionaries of abbreviations, as well as lists of aeronautical, environmental, agricultural, horticultural, biological, electronic, psychological terms, and so on.

65. Also known as an *exhaustive search*.

66. Based on probability theory.

are only 10,000 possible combinations (i.e., from 0000 to 9999). With the aid of a software application, this could be completed in a matter of minutes.

As an aside, this is the reasoning behind the almost universal recommendation that Internet-based accounts should comprise a passcode containing: 1) one uppercase letter; 2) one lowercase letter; 3) one number; 4) one special character; and 5) at be least eight characters long.

The drawback is where there are many possible keys—as was the case with the Enigma cipher—then the computational time could take so long that the message's content may be known by the opposition's actions, rather than through any attempt to crack it. Yet, if the cipher is in the category of, say, a Caesar cipher, then a brute-force search will almost certainly be successful.

Brute-force attacks comprise: 1) positive brute-force attacks; and 2) reverse brute-force attacks. In an attack using brute-force, many passwords are targeted against a single username, or encrypted file.

In a reverse brute-force attack, many usernames are tested with a single password. How would this work in practice? Think of a system that has millions of users, say, a social media platform. There is a high probability that numerous users have chosen the same password because they have *not* followed security tradecraft.[67] Using a reverse brute-force attack, the attacker uses a commonly used password to probe a long-list of user accounts (or encrypted files).

The next type of attack is where some or all the plaintext is known (termed a *known-plaintext attack*). This could be

---

67. Because they have used dates, names, or places that are important to them, which can be obtained from the Internet or other public records. Or, they use common strings of letters or numbers such as *123456* or *qwerty*.

the case where previous messages were deciphered, or parts of the ciphertext were known because the targeted ciphertext was considered to be using the same key (or method). This was the situation with many of the Enigma intercepts because the Nazis used the same plaintext word or phrase, usually at the beginning of each message,[68] so analysts assumed that the word(s) was present in subsequent cryptograms and worked back from there to solve the key.

In the realm of counterespionage, where one's spy apparatus is placed in contact with the opposition's, there arises the possibility of obtaining the adversary's key by way of a *chosen-plaintext attack.* This method calls for the opposition to transmit a message that contains word(s), phrases, or terms chosen by the intercepting party. In this way, the intercept station knows the plaintext to start the analytic process.

Simple in concept, the operational methods of getting an adversary to do this is anything but straightforward. It takes a cunning entrapment, which is the hallmark of espionage novels. Why? Because one cannot hand a message to a foe and ask them to encrypt it and transmit it. The adversary needs to be deceived, so they think they are passing on important information. Like a magician, this is done by setting up an illusion where the audience—i.e., the opponent—is duped.

Take the case where, during the Second World War, the American's became aware that the Japanese were planning a major attack, but they did not know where—all they had was the enemy code name *AF* for that location. So U.S. counterintelligence set up the fraud by transmitting a message in the clear advising that forces on Midway Atoll

---

68. This method became known as a *crib*. Michael Smith, *The Debs of Bletchley Park and Other Stories* (London: Aurum Press, 2015), pp. 52, 150–151.

were running out of freshwater—"We have only enough water for two weeks. Please supply us immediately."[69] If Midway was the target, then this information would be vital to the Japanese military that was planning an invasion.

When the American's intercepted a cryptogram, they reversed engineered the cipher using the plaintext words or terms, and the key was discovered. The target was Midway, and the Japanese were advising attacking forces of the water shortage. The target ("AF") was deduced by reference to water shortage, but what if the entire key was not revealed.

In such cases, a variant known as the *adaptive-plaintext attack* could be attempted. This, however, requires additional deceptive contacts with the opposition to, hopefully, provide further insights into the key's construction. A chosen-plaintext attack is difficult to carry out, but an adaptive-plaintext attack is more complex. It risks that the opposition may become suspicious and could turn the illusion back on the attacker (in a *double-cross* operation[70]).

Consider this situation: say, a friendly military unit or spy is sent into a denied area with the agency's knowledge that the soldiers or spy will be captured. Their classified information or equipment (e.g., encrypted radio communications equipment) will then be analyzed by the opposition's intelligence service.

Once the friendly force understands that their codebooks or enciphering equipment are being used by the opposition

---

69. Edwin T. Layton, Roger Pineau, and John Costello, *And I Was There, Pearl Harbor and Midway—Breaking the Secrets* (New York: William Morrow and Company, 1985), p. 422.

70. For an in-depth discussion of how such counterespionage operations work, see, J.C. Masterman, *The Double-Cross System in the War of 1939 to 1945* (Canberra: Australian National University Press, 1972).

to intercept their communications, false and misleading information can be transmitted to real or fictitious units. The purpose is to have the opposition chase dead ends or pursue moles who do not exist, or better still, accuse "good officers" in the opponent's ranks as being moles, thus ruining their careers and making the opposition self-destructive.

If a code is broken, or if information is obtained from a spy—for example, an agent in place—the agency must be guarded about how it actions this information. Because it could allow the opposition to realize that there is a leak and either change encryption or codes or hunt for the mole. So, the techniques are to suggest to the adversary that the source of this vital information is other than the broken code or the mole. In this way, the agency can continue to amass details of the opposition, and then finally, deliver a decisive blow as in winning the war rather than just the battle.[71]

This anatomical part of intelligence work is termed *counterespionage*[72] because once one engages in the enterprise, it becomes "a wilderness of mirrors."[73] Take the

---

71. An example of this was the November 14, 1940 Nazi raid on the industrial area of Coventry in England. Winston Churchill received intelligence from deciphered Enigma intercepts advising that a large air raid was to take place. This placed Churchill in an invidious position of having to decide to save the city's population and expose the Brits were reading Enigma or protect the code's source. He decided on the later. Anthony Cave Brown, *Bodyguard of Lies* (London: W.H. Allen & Company, 1976), pp. 38–44.

72. Hank Prunckun, *Methods of Inquiry for Intelligence Analysis, Third Edition* (Lanham, MD: Rowman & Littlefield, 2019), p. 23.

73. This is a metaphor for a state-of-mind where it is difficult to distinguish between reality and illusion. Hank Prunckun,

case of the Soviet spy, Gordon Arnold Lonsdale.[74] MI5 was able to break-in to Lonsdale's premises and copied his one-time pads (the numbers on the pads' pages were to be used once and then immediately discarded). MI5 then returned the pads.

Lonsdale was unaware of the theft, so he continued to receive and send messages using these cipher keys. But, having copied the keys, British intelligence was able to decrypt the coded traffic.[75]

So far, we have discussed mathematical methods of decryption, but there are others that exploit the weaknesses of those using the system—*pretext attacks*. These are often termed *social engineering attacks*, but that term is a misnomer.[76] There are many ways information can be revealed, unknowingly, to someone who has used a pretext.[77]

Counterintelligence Theory and Practice (Lanham, MD: Rowman & Littlefield, 2011), p. 9.

74. Lonsdale was the cover name for Konon Trofimovich Molody. Molody posed as a Canadian businessman while he operated the so-called Portland Spy Ring in England from the late-1950s until his arrest in 1961.

75. Robert Reynard, *Secret Code Breaker II: Cryptanalyst's Handbook* (Jacksonville, FL: Smith & Daniel Marketing, 1997), p. 85.

76. The term *social engineering* is often used for this type of activity, but it is "vastly different from its true meaning, which is large-scale societal planning." Hank Prunckun, *Counterintelligence Theory and Practice, Second Edition* (Lanham, MD: Rowman & Littlefield, 2019), p. 131.

77. Most notably used by private investigators. For instance, see, Greg Hauser, *Pretext Manual* (Austin, TX: Thomas Investigative Publications, 1994).

There are so many tips, tricks, and techniques that it is beyond the scope of this book to even summarize them. Suffice to say, that pretexts exploit people's strengths and weaknesses. One might say that targeting a person's weaknesses makes sense—their laziness (comforts), fears (blackmail), or greed (bribes), are all logical vulnerabilities, but their strengths? Yes, for example, their sense of loyalty (a strength) can be played against them by posing, say, as another very loyal person who is in difficulty. The attacker using a pretext, falsely claims that if the "mark"[78] allows them access to the system, they will be saved. Being loyal, the mark allows this.

In sum, for every cryptosystem that is developed, it is a sure bet that an adversary will try to engineer a mathematical method to reveal the key. Or, the opposition may resort to a penetration operation that exploits the system's human interface. Does that mean that all cryptosystems are insecure? No, there are systems that are unbreakable.

---

78. This term is slang for the target of a ruse. George Hayduke, *Screw Unto Others: Revenge Tactics for All Occasions* (Boulder, CO: Paladin Press, 1987), p. 7.

— CHAPTER FIVE —

QUEST FOR THE UNBREAKABLE

Francis Bacon's view of perfect ciphers was: "...that they be not laborious to write and read; that they be impossible to decipher; and, in some cases, they be without suspicion."[79]

Why did he think this? Because cryptography serves three purposes: confidentially, integrity, and authentication. It does this by developing systems that ensure cryptographic traffic remains secret (confidentiality); the messages arrive in the same state as they were sent, and that no one else could have sent it (integrity); and the sender is who they claim to be (authentication).

If the value of the information being sent is viewed as a prized catch, then it is axiomatic that an amount of effort equal to the data's value will be put into cracking the cipher. This raises the question, "Is there an unbreakable cipher"?

The short answer is, "yes." This, however, needs to be qualified—so, here is the long answer. It is reasonable to say that from the time ciphers were first used, its goal was to be easy to write (encipher); hence it was easy to read (decipher). Yet, an adversary was unable to crack the algorithm. This held until someone was able to crack the cipher, so another method was developed.

A very strong cipher based on polyalphabetic substitution was created in 1585—the Vigenère cipher. It

---

79. Laurence Dwight Smith, *Cryptography: The Science of Secret Writing* (New York: Dover Publications, 1955), iv, referring to section XVI (6) of Francis Bacon's book published in 1605, titled *The Advancement of Learning.*

remained an uncrackable cipher for over two-hundred years and was portrayed as *le chiffre indéchiffrable* (the indecipherable cipher). Then, in 1863, the German cryptographer, Major Friedrich Wilhelm Kasiski, devised a method that revealed the key. What was once mathematically unbreakable, was no longer.

We could follow the development from this-cipher-to-that as history progressed, discussing how one cipher replaced another because of what was for a time a security technique. So, let us advance to the time of this writing where the state-of-the-art in encryption has progressed to the point where, mathematically, there is still the possibility of cracking a polyalphabetic substitution cipher, but the time it would take, in a practical sense,[80] makes it unsolvable.

Perhaps the quest for a "perfect" cryptological unbreakable cipher has been reached? This not because of human mathematical prowess but only because we have exhausted our technology to carry out the number of computations in the timeframe that is necessary to retain the value of the data. By making the decryption process so time-consuming and expensive, the rewards no longer outweigh the effort.

This goes some way to explain the reason why other methods are used *in parallel* with cryptological approaches—espionage and counterespionage. This is because it may be easier to obtain a key or codebook by stealth, or cajole, bribe, or blackmail an opposition staffer into handing over the solution than it is to solve it cryptanalytically.

---

80. Not only in terms of time, but in the amount of computer resources that would need to be devoted to the project as well as the costs of financing such an endeavor.

Let us extrapolate—if there are ciphers that have been developed that would take centuries to crack, what methodology was involved in their development? Given the first substitution ciphers were, arguably, developed by the Egyptians some 4,000 years ago,[81] we might be expecting some fantastic calculational cunning that has led to a breakthrough that has changed the way cryptography operates.

This would be nice, but it is not the case. As aircraft design has evolved from what was first designed by the likes of the Wright Brothers, an aircraft still looks the same over one-hundred-years later. The only difference is that rather than having two wings (biplanes), aircraft have a single wing (monoplanes). Granted, there has been prodigious improvements in propulsion, aerodynamics, metallurgy, and avionics, yet the configuration of a plane still looks the same.

This example is analogous to the development of ciphers. There has not been a transformation—or should it be a *metamorphosis*—from the Caesar "caterpillar" cipher to the equivalent of a "butterfly" cipher. Like aircraft, the same mathematical principles that applied in past epochs are still relevant today.

As a wing provides lift for an airplane, the "lift" provided to a cipher is its central tenet—make it difficult to solve. Rather than move from two wings to one, ciphers have grown from monoalphabetic ciphers[82] to cryptograms based

---

81. David Khan, *The Code-Breakers: The Story of Secret Writing* (Toronto: The Macmillan Company, 1967), p. 71.

82. Monoalphabetic ciphers use a single fixed substitutional shift for encrypting a plaintext message. Compare this to *polyalphabetic* ciphers that uses more than one substitutional shifts for a plaintext character, which presents multiple possibilities in the cyphertext.

on combinations and permutations so enormous in possibilities that intuition and pencil-and-paper no longer suffice, nor does using a supercomputer.

It could be said that the unbreakable ciphers of today are still caterpillars, but their size is so big no one can "step" on them. Although this is likely to be true for now, what developments may cause a change in this situation?

The answer that immediately presents itself—other than having recruited a mole in the opposition's code room—is computational power. This suggests *quantum computing.*

Let us look at an example of how such a situation might play out. For our purpose, we will take a reasonably strong cipher that uses a 256-binary bit key. Recall, a binary bit is either a 0 or a 1. Therefore, arithmetically, a 256-bit key would provide 2^{256} possible key combinations. Nevertheless, the key may be found after trying half of all the possible combinations[83] so that this number would be 2^{255}.

Assume that we have secret intelligence that advises us that the opposition has a built a computer that comprises one-hundred parallel GPUs[84] that can affect a brute-force attack at the rate of one-billion keys per second, per GPU. This number can be expressed as 100×10^9 possible key combinations per second.

To calculate the time our notional adversary's computer would take to go through each possibility, multiply the number of seconds in a year (i.e., $60 \times 60 \times 24 \times 365.25 = 31{,}557{,}600$) by the number of computations per second (i.e., 100×10^9) to determine the total number of keys that can be tested yearly (i.e., $31{,}557{,}600 \times 10^{11} =$

83. See page 39 for an explanation.

84. *Graphics processing units* are electronic devices fitted to computers that accelerate computations.

31,557,600,000,000,000,000). We see that the opposition's computer can process approximately 31.56 quintillion keys per year.

On the face of this result, we might conclude that the problem might be solved in minutes. Yet, the opponent is trying to crack a 2^{255} key cipher. The possibilities are much greater!

To get a sense of how long it would take, 2^{255} possibilities is divided by 31,557,600,000,000,000,000 computations. The answer is 6^{235} years. Given that the age of the Earth is around five-billion years (i.e., 5,000,000,000, or 5 x 10^8), it is beyond the capability of the opposition to crack to cipher.

Even if their computing power was doubled, it would still take 6^{234} years. It would be a compelling argument to posit that even if a supercomputer was tasked to solve the problem, it could not be done in less than a million years.[85]

Moreover, this is a hypothetical calculation that does not consider the cost of the resources needed to run the algorithm. Take, for instance, just the cost of electricity. Even if our notional opponent had a supercomputer dedicated solely to the task of decrypting a single 256-bit ciphertext, the computer would likely consume about four-megawatts to run. Four-megawatts purchased at \$0.10/kWh would equal \$400 an hour, or in a yearly figure, about \$3.5 million.

What will happen when quantum computers are perfected? It has been hypothesized that these devices will sound the death knell for encryption. Perhaps, but if keys

---

85. It may even be longer—much longer—but that calculation is beyond the scope of this book.

are maintained at 256 and larger, then the law of diminishing returns may still be the deciding factor.[86]

Even so, there are other variables that might influence whether a cryptanalyst succeeds or fails, and that involves the advances being made in cryptographic algorithms. These developments are aimed at making searching faster.[87] Performed on a quantum computer, Shor- and Grover-type algorithms could reduce the time it takes to crack a cipher. Could they reduce it by centuries? We could speculate, but until cryptographers have the opportunity to experiment with quantum computers, all remains conjecture.

What is not conjecture is the interest the three great power blocs, other state-actors, and powerful criminal syndicates have in archiving cryptological messages being transmitted now for when, one day, there exists software and quantum hardware that can decrypt them in days or weeks, rather than millennia. These historical cryptograms could be exploited to yield very profitable information.

> The Dutch General Intelligence and Security Service singled out a looming threat that adds even more urgency to the need for quantum-safe encryption. In a scenario it calls 'intercept now, decrypt later,' a nefarious attacker could start intercepting and storing financial transactions, personal emails and other sensitive encrypted traffic, and

---

86. John F. Dooley, *History of Cryptography and Cryptanalysis: Codes, Ciphers and Algorithms*, p. 258.

87. For instance, *Shor's algorithm* and *Grover's search algorithm*. See, Peter Shor, "Polynomial-Time Algorithms for Prime Factorization and Discrete Logarithms on a Quantum Computer," in *Proceedings of the 35th Annual Symposium on Foundations of Computer Science*, Santa Fe, IEEE Computer Society Press, 1994, pp. 124–134; and Lov Grover, "A Fast Quantum Mechanical Algorithm for Database Search," in *Proceedings, 28th Annual ACM Symposium on the Theory of Computing (STOC)*, Philadelphia, 1996, pp. 212–219.

> then unscramble it all once a quantum computer becomes available.[88]

It does not take a great deal of imagination to envision situations where such information could be used for extortion, intimidation, corruption, and coercion against heads of state, public officials, aid organizations, and leaders of commerce and industry.

---

88. Chris Cesare, "Encryption Faces Quantum Foe," in *Nature*, September 9, 2015, p. 167.

— CHAPTER SIX —

AN UNBREAKABLE CIPHER

With the inevitable commercialization of quantum computing and the development of clever search algorithms, we could become despondent about cryptography's ability to keep information secret. Yet, there are, based on mathematical principles, ciphers that cannot be defeated by brute-force attacks.

One such cipher uses a *one-time pad* (OTP). This method is based on the rule that for every plaintext character there is a corresponding ciphertext key. Unlike the Caesar cipher where the key shifts a specified number of characters left or right, a one-time pad encrypted message of, say, 250 characters, or bits, would have 250 keys.

If a brute-force attack were mounted, the result would be 250 strings of possible characters—including the correct key string—but it would not be possible to determine the correct string. Why? Because each plaintext character is subjected to a random shift,[89] and that the key is always the same length as the message. This outcome is a shift that never forms a pattern because the distribution is uniform—that is, it is random. This makes the one-time pad uncrackable.[90]

---

89. There are many commercial hardware devices that can generate random numbers. These devices use naturally occurring environmental noise to produce the random numbers (as opposed to devices or methods that result in pseudorandom numbers that only approximate randomness).

90. Alan Stripp, *Code Breaker in the Far East* (Oxford: Oxford University Press, 1995), p. 100.

The OTP cryptosystem was developed by Frank Miller in 1882. Miller was a banker in California and devised the system to ensure secrecy for commercial arrangements and transactions.[91] During the Second World War, the British Special Operations Executive (SOE) and the U.S. Office of Strategic Services (OSS) used one-time pads and the teleprinter hotline between Washington DC and Moscow, which was installed after the Cuban Missile Crisis, used a commercial one-time tape system.

Oxymoron – Random Order

To use a one-time pad, there are a few procedural requirements. Matching pads (or other physical or electronic forms) need to be produced and distributed. There also needs to be an agreed sequence of use between the sender and receiver for each sheet (if produced in a printed pad) or another method if produced in an electronic or alternative format.[92]

The sender—"Alice"—uses the random sequence of numbers on the agreed sheet to encipher her message. When the receiver—"Bob"—collects the message, he uses the matching sheet of identical numbers to decipher it.[93]

---

91. Steven M. Bellovin, "Frank Miller: Inventor of the One-Time Pad," in *Cryptologia*, Volume 35, Number3, 2011, pp. 203–222.

92. The usual method for doing this was to arrange that a specific sheet was to be used on a specified date or that sheets were to be used in a particular sequence, etc. For instance, "on January 4, use the fifth sheet," or "for the next message, use the subsequent sheet."

93. Rather than use "A" and "B" as sender and receiver, in 1978 cryptographers Rivest, Shamir, and Adleman created the fictional characters Alice and Bob. Since then, other characters have been added to the lexicon—such as "Eve," who is an eavesdropper.

The most common, but not exclusive, technique for performing the encryption/decryption tasks is to first assign each alphabetic letter a number—"A" is 0, "B" is 1, "C" is 2, "D" is 3, and so on.

The next step is to combine the message with the key using *modular addition*. In the case of the Roman alphabet, this is *modulo 26*.[94] To demonstrate this, let us look at the plaintext message "LIBRARY." The length of the key to encipher it needs to be the same length—"PXBMKGC"—and combined as shown in Table 1.[95]

When the sum of the number representing the plaintext message and the number representing the key are greater than 26, then, in modular arithmetic, we continue counting by wrapping around Z by starting over at the letter A.

See, Ron L. Rivest, Ron L.; Adi Shamir; and Lenard Adleman, "A Method for Obtaining Digital Signatures and Public-Key Cryptosystems," in *Communications of the ACM*, Volume 21, Number 2, February 1978, pp. 120–126.

94. Modular arithmetic is a system for adding numbers (i.e., integers) that allows the numbers to "wrap around" when they reach a set value (known as the *modulo*). A twelve-hour clock uses modulo 12. For example, to calculate four hours from 10am, we start the process of adding but when we reach 12noon, we "wrap around" and start back a "1," arriving at our answer 2pm. Hence, modular arithmetic is informally referred to as *clock arithmetic*.

95. This can also be done using the Fibonacci system were, when adding numbers, numbers greater than 9 are not carried forward (i.e., modulo 10). Take this as an example, 3,765 and 1,196 equal 4,851 using the Fibonacci system. When deciphering, because "carrying" was not allowed, "borrowing" is permitted while subtracting. Peter Wright with Paul Greengrass, *Spy Catcher: The Candid Autobiography of a Senior Intelligence Officer* (Richmond, Victoria: William Heinemann Australia, 1987), p. 179.

After Alice sends her ciphertext ZGDEMYB to Bob, he uses the matching key on the corresponding page of his pad, in reverse, to decipher the message. "Eve" may intercept the message, but she cannot decrypt it because this method is *information-theoretically secure.*[96] That is, the ciphertext provides no information—e.g., patterns—about the plaintext message.

Plaintext message	L	I	B	R	A	R	Y
Assigned number	13(L)	9(I)	2(B)	18(R)	2(B)	18(R)	25(Y)
Add key	16(P)	24(X)	2(B)	13(M)	11(K)	7(G)	3(C)
Equals	26	33	4	31	13	25	28
Mod 26	26(Z)	7(G)	4(D)	5€	13(M)	25(Y)	2(B)
Ciphertext	Z	G	D	E	M	Y	B

Table 1—One-time pad procedure.

To ensure the key remains safe, both Alice and Bob immediately destroy the sheet from the one-time pad.[97]

---

96. Based on information theory, this term refers to a cryptosystem that cannot be broken even if the opposition possesses an infinite amount of computing resources. According to the theory, this type of system is cryptanalytically unbreakable because intercepted message traffic does not contain enough information to provide an attack vector. Claude E. Shannon, "Communication Theory of Secrecy Systems," in *Bell System Technical Journal*, Volume 28, Number 4, October 1949, pp. 656–715.

97. It is understood that intelligence agencies have been known to issue its agents one-time pads that were printed on "flash paper." This material comprised a chemically treated paper in the form of nitrocellulose. This material burned instantaneously, leaving virtually no ash. Organised criminals, such a "bookies," were known to use *potato paper*, which dissolved in water (e.g., dropping it into a cup of coffee). John Douglas and John

The classic espionage one-time pad is typified by the example displayed in Figure 13. With technology advancements since the days of SOE and OSS, espionage agents are almost certainly now using computer technology with messages being transmitted by way of the Internet. Though one-way radio voice traffic from numbers stations still occurs.

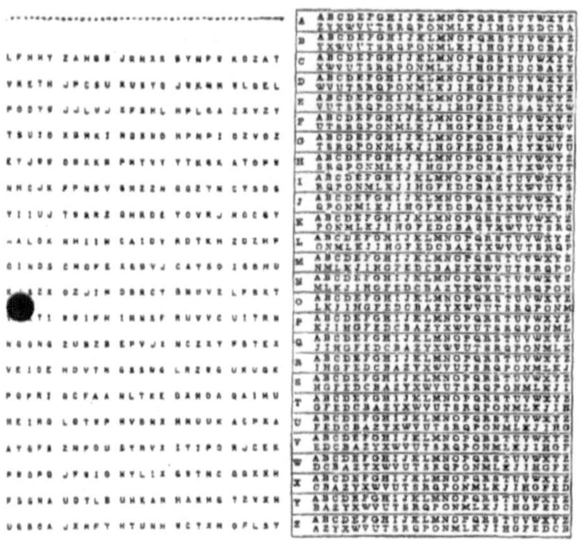

Figure 13—An example of a CIA DIANA class one-time pad.[98]

As a declassified example, take the case of Cuban agent Ana Belen Montes, who operated in the United States in the early-2000s. She received encrypted voice messages on a battery-operated portable Sony shortwave radio from numbers stations that were based in Cuba. She then entered the numbers into a text document she created on her laptop

---

Olshaker, *Mind Hunter: Inside the FBI's Elite Serial Crime Unit* (New York: Gallery Books, 1995), p. 61.

98. Source: "A History of U.S. Communications Security," in the *David G. Boak Lectures* (Fort George G. Meade, Maryland: National Security Agency, 1973), p. 22.

computer. Using a key on a removable diskette (rather than a pad), she ran a program that deciphered message in the same way Bob would have done so using pencil-and-paper in the example shown in Table 1.

However, rather than following sound counterintelligence processes by immediately destroying the messages and keys, she left them on the laptop's hard disk drive.[99] When the FBI searched her apartment, they were able to recover this incriminating evidence. Even though the software one-time pad cipher was not broken, it was the spy's inept tradecraft that was revealed the secrets.[100]

Defeating a one-time pad cryptosystem have been attempted. The notable U.S. counterintelligence Venona project was one. This was a secret program that ran for thirty-seven years—from February1, 1943 to October 1, 1980. It has been reported that during this time:

> There were about 3,000 Venona messages translated. Partial information was available from many messages as early as 1947and later that year was provided to the FBI. The Venona translations now released to the public often show an unexpectedly recent date of translation because the breaking of strong cryptographic systems is an iterative process requiring trial and error and reapplication of new discoveries, leading to additional ones. A message may have been reworked many times over the years as new discoveries enabled progress in the decryption and

---

99. Special software is needed to "wipe" a hard disk drive. This is because using the "delete" function or "format" command will only delete the file indexing system. This makes the data invisible, but it is still present on the hard disk drive. These data can be recovered using a file recovery program.

100 Robert Wallace and H. Keith Melton, with Henry R. Schlesinger, *Spycraft: The Secret History of the CIA's Spytechs, from Communism to al-Qaeda* (New York: Dutton, 2008), p. 452.

> understanding of more and more of the text. Only the last and best translations have been released. Almost all of the KGB messages between Moscow and New York and Moscow and Washington of 1944 and 1945 that could be broken at all were broken, to a greater or lesser degree, between1947 and 1952.[101]

Cracking a one-time pad cipher is more successful if errors are discovered in the opposition's implementation. Areas that are exploited include probing to see if the keypads are genuinely random; if not, this becomes an attack vector.

Mistakes made by cipher clerks and agents in distribution and protecting the keypads, errors in the production of large numbers of random numbers (i.e., quality control) and printing (or creating electronic files), as well as the covert obtaining of keypads, are possible opportunities that are part of the "…iterative process [of] trial and error and reapplication of new discoveries…"

There have been many improvements, versions, and new designs to one-time pads. These have developed to accommodate electronic means of communicating. To demonstrate how such progressions take place, we will look at how an older analogue technique (i.e., DIANA) moved on to a more efficient version (i.e., the ORION variation), consider this example:

> To understand how the ORION pad is used, it will be helpful to visualize the two illustrations shown above as being printed in exact alignment on reverse sides of the same sheet of paper. To encipher, one sheet of the pad with the straight alphabet side up is placed on a piece of carbon paper, carbon side up. With this arrangement, when a plaintext letter is circled on one side of the paper, a circle will appear on the other side

---

101. Robert Louis Benson, *The Venona Story* (Fort George G. Meade, Maryland: National Security Agency, Center for Cryptologic History, 2001), pp. 14–15.

of the paper as surrounding the cipher letter because of the carbon paper. Therefore, by recording the text of the message—one letter per line—on the plaintext alphabet, the enciphered text is available by merely turning over the page.[102]

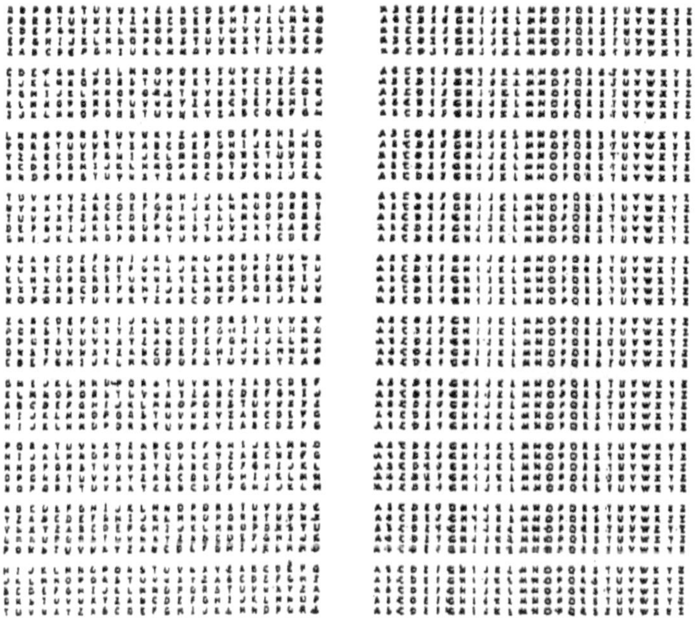

Figure 14—An example of the National Security Agency's ORION one-time pad.

This manual improvement in the speed of enciphering has, in all probability, been replicated digitally using software applications designed for the purpose. If this assumption is correct, such programs would be classified, so review here is not possible.

---

102. U.S. National Security Agency, *A History of U.S. Communications Security (Volumes I and II); the David G. Boak Lectures* (Fort George G. Mead, MD: National Security Agency, 1973), p. 24.

Yet, we can speculate that given that word-processing programs can convert voice dictation to text; then it is not unreasonable to propose that a field agent could speak into her portable device and have her message converted to plaintext. Then, with a few keystrokes (or voice commands), she can have the application translate her message into ciphertext using a one-time digital pad located on, say, a disposal USB dive.

The resulting cryptogram could then be transmitted, not through a "dead-drop,"[103] shortwave transmission,[104] "brush-pass,"[105] or a "cut-out,"[106] but by email or using an FTP-upload protocol that allows file transfer between remote computers, or some other electronic means.

This proposition could be extended to encrypting real-time voice messages. Radiotelephony, as well as landline and wireless telecommunications used voice scramblers. These devices took the sender's voice and randomize (i.e., pseudorandom) the sounds using an algorithm before it is transmitted. At the receiver's end, a matching device

---

103. A dead-drop can be either a place or a repository for intelligence operatives to leave and collect messages or artifacts. It allows operatives to do this without the need to make contact.

104. e.g., see Henry Prunckun, "Covert Radio Communications: A Viable Tactic for International Terrorists"? in *Defense and Security Analysis*, Volume 30, Number. 2, June 2014, pp. 176–184.

105. Also known as a *hand-off*, this is a method of delivering information. It is performed in public but to anyone observing, it looks like two strangers passing. Best performed in crowd places to make it difficult from surveillance operatives to known which passer-by might be the hand-off agent.

106. A cut-out is a person who acts as an intermediary between two operatives in an espionage network. This is done so that the principal operative remains anonymous to the junior member.

reverses this analog process to yield intelligible two-way (full duplex) conversation.

As we know, a fixed algorithm and a pseudorandom key can be broken, so another process is introduced to provide a high-level of security.[107] Nonetheless, the generation of voice scramblers that are on the market at the time this book went to press has moved from analog scramblers to digitizers that are coupled with encryption circuits. They operate by converting the original analog voice signal into digital form, then encrypting these data and sending them. These devices are capable of transmitted information classified as Secret and Top Secret.

These digital enhancements certainly satisfy Francis Bacon's dream of developing a "perfect cipher." They allow information that needs to be protected, to be protected.

---

107. Although such devices may incorporate a random generator that can produce a random seed value.

ABOUT THE AUTHOR

Dr Henry (Hank) Prunckun is a former Australian government intelligence officer, as well as a freelance private investigator, who spent much of his career in various operational fields, including government security, investigation, and counterterrorism. He is the winner of two literature awards and a professional service award from the International Association of Law Enforcement Intelligence Analysts. Dr Prunckun also served for over a decade as a research criminologist at Charles Sturt University, Sydney, specializing in the study of transnational crime—espionage, terrorism, drugs and arms trafficking, and cyber-crime.

INDEX

Bacon, Francis, 47, 47n79, 63
Biuro Szyfrów, 13–14
Brush-pass, 62, 62n105
Brute force, 40, 40n64
Churchill, Winston, 10, 44n71
Cipher
  Caesar, 25–26, 26, 36–37, 41, 49, 54
  perfect, 47, 48, 63
  polyalphabetic, 14, 37–38, 47–48, 49n82
  substitution, 14, 21, 23, 25–29, 26n41, 36–38, 47–49, 49n81, 49n82
  transpositional, 3–4, 13, 21, 23, 27–28
  Vigenère, 37, 38, 47–48
Code
  codebook, 29, 30, 30n48, 31, 32, 34, 43
  defined, 23
  group, 29
  Morse, 11, 15, 23, 24, 25, 26
  number, 23
  phrases, 11, 23, 28, 42
  polyalphabetic, 14, 37–38, 47–48, 49n82, 44, 48
  Q, 23
  word, 23, 29, 31
Counterespionage, 18–19, 42, 43n70, 44–45, 48
Counterintelligence, 18, 18n25, 29–30, 30n47, 30n48, 32, 42–43, 59
dead-drop, 62, 62n103
Enigma, 13–15, 14n16, 15n17, 18n24, 38, 41, 42
FBI, 57–58n97, 59–60
Franklin, Benjamin, 29, 29n45
Great War, 30, 33n52
Harden, Donald and Bettye, 2, 7
Kerkhoff's Doctrine, 39, 39n61
KGB, 60
National Security Agency, 58n98, 61, 61n102
Numbers stations, 18–19, 19n26, 20n31, 58–59
One-time pad, 45, 54, 55, 55n91, 57, 57n97, 58, 59, 60, 61, 62
Pseudorandom numbers, 14, 14n15, 38, 54n89, 62, 63
Random numbers, 54n89, 55, 60
Scytale, 12, 13, 16
Shortwave, 19, 19n27, 20n31, 58, 62, 62n104
SW. *See* shortwave
Venona, 59–60, 60n101
Yardley, Herbert, 13n13, 33
Zodiac, 1–9

www.ingramcontent.com/pod-product-compliance
Ingram Content Group UK Ltd.
Pitfield, Milton Keynes, MK11 3LW, UK
UKHW041632190726
13854UKWH00006B/2458